Energy 108

伴日航行

Sailing with the Sun

Gunter Pauli

[比]冈特·鲍利 著

[哥伦]凯瑟琳娜·巴赫 绘

高 芳 李原原 译

上海遠東出版社

特别感谢以下热心人士对童书工作的支持：

匡志强　宋小华　解　东　厉　云　李　婧　庞英元

李　阳　刘　丹　冯家宝　熊彩虹　罗淑怡　旷　婉

杨　荣　刘学振　何圣霖　廖清州　谭燕宁　王　征

李　杰　韦小宏　欧　亮　陈强林　陈　果　寿颖慧

罗　佳　傅　俊　白永喆　戴　虹

目录

Contents

一只海
狮在看一艘船启
航。一群海鸥正准备加入这次
航行。
“这艘双体船有4个帆，比普通的船多
2个。而且那些帆不是普通的帆。”
海狮说。“显然不是普通的船，
这一定是艘昂贵的船。”

A sea lion is
watching a boat set sail.
A flock of seagulls are getting
ready to join the voyage.
“This catamaran has four sails.
That is two more than usual. And
those are no ordinary sails,”
remarks the sea lion. “This is
clearly no ordinary vessel. It
must be an expensive
boat.”

这一定是艘昂贵的船

It must be an expensive boat

但是你看，它有太阳能电池

But look, it has solar cells

“但是你看，

它有太阳能电池。”海鸥观

察并指出。“我从这上面可以看

到，这艘船甚至利用螺旋桨引起的

湍流产生动力。”他补充说。

“这是怎样实现的呢？”

海狮很好奇。

“But look, it has solar
cells,” observes the seagull.
“The boat even gets power
from the turbulence created by
a propeller. I can see that from
up here,” he adds.

“How does that work?”
wonders the sea lion.

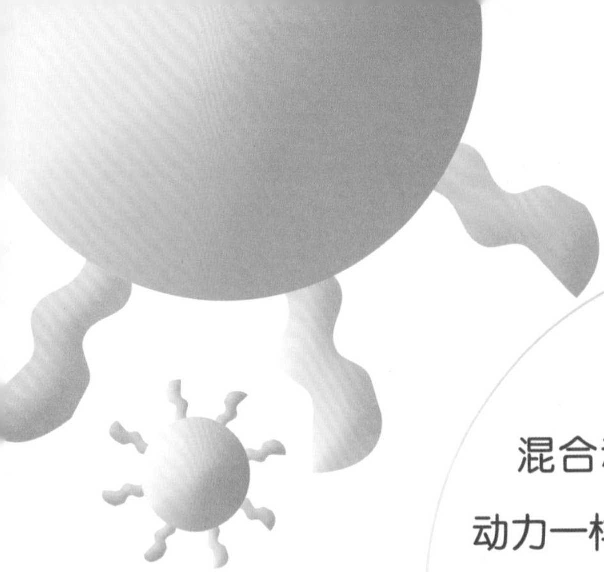

“嗯，就像那些
混合动力汽车利用刹车产生
动力一样，这艘船有一对螺旋桨，
可以把推进力和水流所产生的多余尾
流全部利用起来。”

“这是那种‘绿色’船只中的
一艘吗？”海狮问道。

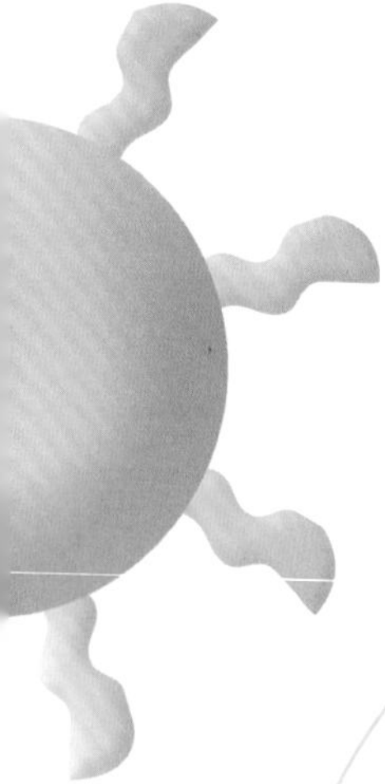

“Well, just like those hybrid cars get power from using their brakes, this boat has a double propeller that catches all the excess wake from propulsion and the current.”

“Is this then one of those ‘green’ boats?” asks the sea lion.

这是那种“绿色”船只中的一艘吗？

Is this one of those “green” boats?

它被称为“蓝色”船只

This is called a “blue” boat

“不，它被称为‘蓝色’船只。毕竟，水是蓝色的，天空是蓝色的，从外太空看地球也是蓝色的。”

“这听起来很浪漫，但我们得面对现实。一艘有4个帆、太阳能电池以及双螺旋桨的船肯定更贵吧？3种能源替代了化石燃料，那些沉重、肮脏的东西才便宜！”海狮惊呼道。

“No, this is called a ‘blue’ boat. After all, the water is blue, the sky is blue, and the earth viewed from outer space is also blue.”

“That sounds very romantic, but let’s be realistic here. A boat with four sails, solar cells and double screws must surely be more expensive? Three sources of power instead of just fuel, that heavy, dirty stuff that is supposedly cheap!” exclaims the sea lion.

“嗯，这种双体船不使用内燃机而使用电动引擎。在无风的时候，帆被展平，内置的太阳能电池就会从早到晚工作。”

“打断一下，太阳能电池不是只在白天工作吗？我从没听说太阳能电池可以在晚上工作。”

“Well, this catamaran does not use a combustion engine but rather an electric engine. And when the wind is calm, the sails are flattened and the built-in solar cells work day and night.”

“Excuse me, but don’t solar cells only work during the day? I’ve never heard of solar cells working in the dark.”

太阳能电池不是只在白天工作吗？

Don't solar cells only work during the day?

我真的不知道

I really don't know

“真的吗？当一个表面漆黑的物体被置于晴朗的夜空下，你认为会发生什么呢？”海鸥问道。

“我真的不知道。”海狮回答。

“Really? What do you think happens when something is black and it’s dark outside, with clear skies?” asks the seagull.

“I really don’t know,” replies the sea lion.

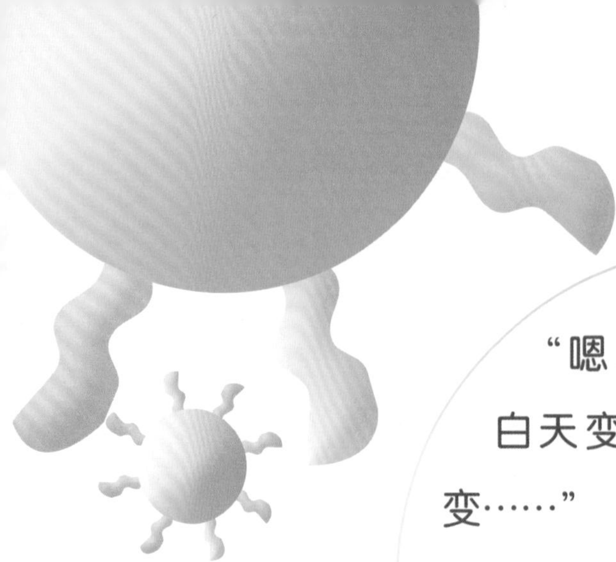

“嗯，试想一下。如果白天变热，那么夜晚它会变……”

“变冷吗？”

“是的。如果你捕捞到很多新鲜的鱼，你会用冷水做什么呢？”

“Well, think about it. If it gets hot during the day, so during the night it will get ...”

“Cold?”

“Indeed. And what can you do with cold water if you have lots of freshly caught fish?”

你会用冷水做什么呢？

What can you do with cold water?

刚捕捞到的鱼需要冷藏

Freshly caught fish need to be kept cool

“嗯，我猜想刚捕捞到的鱼是不是需要冷藏来保持新鲜和美味？”

“完全正确，”海鸥回答，“我们需要帆捕捉风和阳光来发电，需要用冷水保持鱼的新鲜，以及用螺旋桨的湍流来产生额外的电力。这是一艘绝妙的双体船。”

“Well, I suppose freshly caught fish need to be kept cool to stay fresh and tasty?”

“Exactly,” replies the seagull. “We need sails to catch the wind, sunlight to make electricity, cool water to keep the fish fresh, and the turbulence of the propeller to generate additional power. This is a wonderful catamaran.”

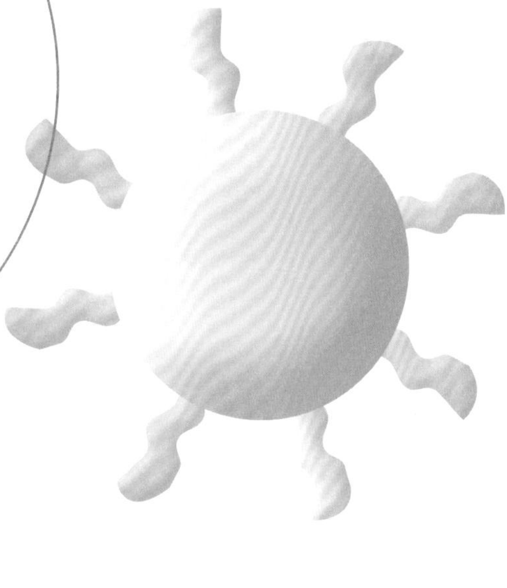

“并且，
如果鱼被冷鲜保
存，可以保持味道鲜美。
嗯……我最好还是和你一起加入
这次航行，而不是直到你返航了
还在岸边徘徊。”
……这仅仅是开始！……

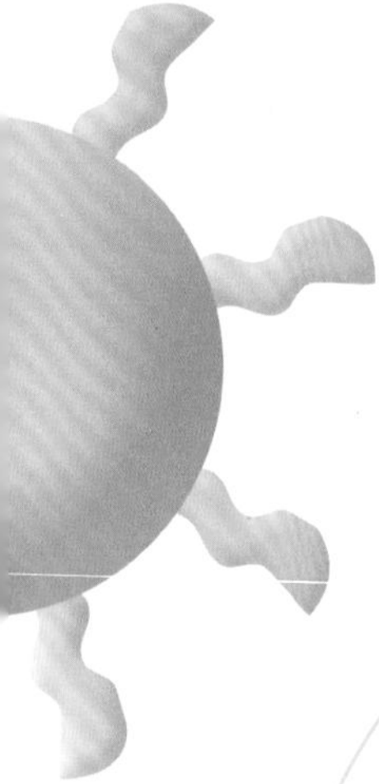

“And, if fish is kept cold
and fresh, it remains so tasty.
Mm … I may as well join you on
this voyage instead of lingering
on the shore until you return.”
… AND IT HAS ONLY JUST
BEGUN!…

……这仅仅是开始！……

… AND IT HAS ONLY JUST BEGUN! …

Did You Know?

你知道吗？

A sailboat is both pushed and pulled forward by the wind. It works on the same aerodynamic principle as that of the wind flow over the wings of an aircraft.

一艘帆船同时被风推动和拉动着向前。它应用的空气动力学原理与风在飞机机翼旁的流动是一样的。

Generating power by braking, as first used in cars manufactured by Toyota and Honda, was already invented during the 1890s and applied to the electric railway.

19世纪90年代再生制动被发明，它首次应用于丰田和本田的汽车制造，现已被应用于电气化铁路。

The catamaran has two hulls and was first used by the Tamils 6 000 years ago. The term “catamaran” is derived from the Tamil word *kattumaram*.

双体船有两个船体，在6 000年前被泰米尔人首次使用。“双体”一词来源于泰米尔语kattumaram。

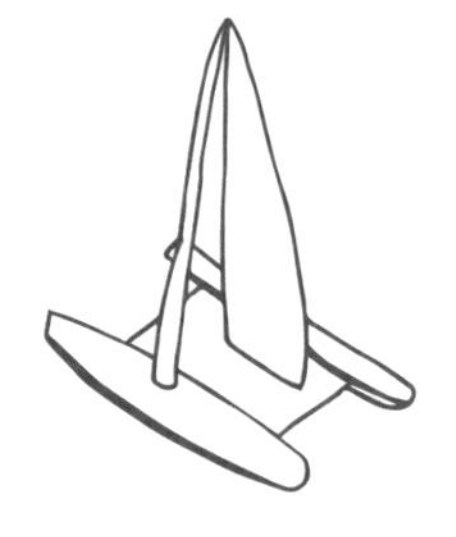

When there is only one power source, the risk of power failure is high. However, when there are three different sources of power, and renewable ones at that, the chance of continuous access to power is high.

当只有一个电源时，电源故障的风险很大。然而，当有三种不同的电力来源，且有一种是可再生能源时，持续获得电力的概率就会很高。

方尖碑是高高的四面锥形的纪念碑，它在公元前200年的古埃及被建造时，由双体船在尼罗河上运载。

The obelisks, the tall, four-sided, tapered monuments originally made in Ancient Egypt, were transported on the Nile in catamarans in 200 BC.

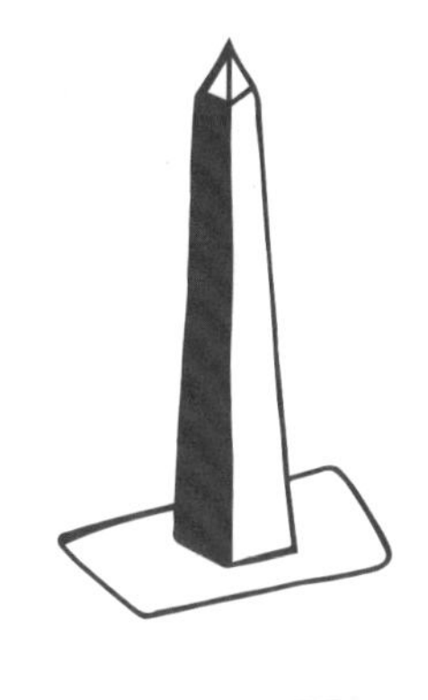

如今世界上有400万艘渔船，只使用桨和帆的有180万艘，而其他渔船，即超过50%的渔船都有引擎。印度尼西亚有70万艘渔船，其中一半没有引擎。

There are 4 million fishing boats in the world today. 1.8 million of these use only oars or sails, whereas the others – just over 50% of them – have an engine. Indonesia alone has 700 000 fishing boats, half of which are without an engine.

在公元前3000年，埃及人制作棉质的帆，以便利用风来协助桨的推进。风帆和人力（划桨）的组合使埃及人在地中海航行成为可能。

In 3 000 BC, the Egyptians made cotton sails to use the wind to assist propulsion by oars. The combination of sails and manpower (to pull the oars) allowed the Egyptians to sail the Mediterranean.

黑冰，也被称为透明冰（一层薄薄的冰，我们能看见在它下面的黑色道路或表面）。当物体在空旷的夜空下散发热量时，黑冰在物体表面生成，甚至在环境温度不足以冻结时也能产生。

Black ice, also called clear ice, (a thin layer of ice that allows the black road or surface underneath it to show through) is generated on surfaces when the open night sky radiates heat. This ice is produced even when the ambient temperature is not freezing.

Do you think solar panels can work at night?

你认为太阳能电池板能在晚上工作吗?

Are green products in general and green energy in particular more expensive than conventional products or power?

绿色产品尤其是绿色能源比传统产品和能源更昂贵吗?

Should the government subsidise green products and power to make them cheaper and more competitive?

政府应该补贴绿色产品和能源使它们更便宜、更有竞争力吗?

Is the price of petroleum the only cost involved in using this kind of fuel? Or are there any hidden costs?

石油的价格是使用这种燃料的唯一成本吗?还是存在其他隐性成本?

Have you ever gone sailing? If you live too far from the sea or a lake, study sailing with a miniature boat in the sink or bathtub. Pay special attention to how the boat manages to sail against the wind. This is an interesting technique that everyone should master: Using the force against us to go forward. This does not only apply to sailing, but is also a good life lesson.

你曾经航海过吗？如果你住的地方离海或湖太远，那就在水槽或浴缸里研究一下小型船的航行吧，特别注意如何操控船逆风航行。这是一项很有趣的技术，每个人都应该掌握：利用本该阻碍我们的力推动我们前进。这不仅适用于航行，也是一条很好的生活经验。

学科知识

Academic Knowledge

生物学	海狮和海豹的区别。
化　学	沸石和熔盐类的化学物质吸收和释放热量比较快，在低温下储存可以避免不必要的化学反应。
物　理	当帆和风的方向相同时，它推着船前进，当逆风航行时，帆拉动船，即把空气引向船的后部并加速；帆的一面高压强，另一面低压强，压强差使帆获得推力，正如空气流过机翼上下的方式；辐射冷却靠的是将热量辐射到外部空间。
工程学	再生制动是一种能量回收机制，在一辆车减速时，它的动能被转化为可以立即使用或储存的能量，这种技术最近被用于混合动力汽车；宇宙飞船在真空中穿梭，靠辐射散发多余的热量；屋顶降温系统结合了高光学反射率和高红外发射率，白天减少太阳的热传递，晚上靠辐射增强散热；智能电网使我们能协调多种能源持续供电。
经济学	我们面临的一个问题是绿色经济作为一种潜在的理想生活方式，维护起来是非常昂贵的；所谓廉价化石燃料的隐性成本：尽管最初看起来很便宜，但它涉及许多环境成本和对我们健康造成的危害，这些都是我们所说的转移成本，即成本由社会承担；船靠3种可再生能源获取动力更加昂贵，但由于不需要化石燃料，也是值得的。
伦理学	我们怎样才能维持绿色经济？它是如此昂贵的可持续绿色生活方式以至于只有富人才能承担得起。
历　史	波利尼西亚人以及德拉威人穿越太平洋航行时首次使用双体船；1886年，弗兰克·斯普拉格在美国首次使用再生制动。
地　理	从外太空看到的地球是蓝色的，伴有一些白色、绿色和棕色斑点；在大海中的航行告诉我们，两点之间最短的旅行方式不在一条直线上，首先因为地球是圆的，其次因为水流和风使船选择沿着阻力最小的路径航行。
数　学	安德烈·柯尔莫哥洛夫的湍流能量定律；翼帆的几何设计提供了比传统帆更大的升力；翼或帆顶部和底部的曲面是不对称的，这能帮助产生空气动力。
生活方式	现代社会根据需要采用开关简单地开关电能，而大自然的能量系统要更复杂，但仍然可以控制；人们倾向于简单和传统的生活方式，但那是不可持续的，而可持续的生活方式被认为是复杂的。
社会学	陪同他人航行的乐趣。
心理学	不愿相信一些事情，即使它被解释得很好也能举出例子；利用逆着我们的力来前进。
系统论	用多种能源产生能量，减少能源需求，使用可再生能源，有可能让所有人类活动自给自足。

情感智慧

Emotional Intelligence

海　狮

海狮正在观察分析新的帆船。他注意到太阳能电池内置在坚硬的帆里，从而得出结论：这想必是一艘昂贵的船。他意识到自己的无知，但很好奇并有勇气提问、探索，并试着去了解，然后得出结论：绿色或蓝色技术听起来浪漫，但一定很昂贵。他并不相信太阳能电池可以在晚上工作，但也承认，他并不真正了解。他聆听海鸥的解释，并产生了加入海鸥航行的动力，享受海鸥的陪伴和美味的食物。

海　鸥

海鸥准备解释一切。他观察和分析船所使用的不同能源的区别，相当了解一切是如何运转的。他有清晰的观点并愿意分享意见。他意识到自己的一些见解不容易理解时就进行解释，通过提问来帮助海狮看透本质。海狮竭力使用常识时，海鸥坚持尝试，并通过提示来建立海狮的自信，否则他可能会自卑。这创造了一个二者的组合，海鸥也成功激励了海狮加入他的钓鱼之旅。

艺术

The Arts

你如何表现湍流？要求学习者每次使用不同风格的图画表现湍流。每个人的图画风格不同，湍流的表现手法也会不同。现在看看这些不同的图画，选择其中最成功地表现湍流的。画幅图来说明如何能够获取湍流的能量而不浪费。

思维拓展

Systems: Making the Connections

现代社会大多数家庭家里都通电，却不知道电是从哪来的。现在电力已经能够进行远距离传输，而能量的产生已经成为经济进步和发展的动力，但当前的系统是不可持续的。碳排放达到一定水平，就会使空气中集聚大量的有害颗粒物，导致气候变化，增加健康风险。

向可再生能源发展的转换进展十分缓慢，而且大多数未经测试。太阳能电池板放置在屋顶，风车放置于山脊，消化器里产沼气，这些能源混合后输送到现有的输电网，是对传统能源，即煤炭、石油、核能和天然气的补充。

船的设计集成了多种可再生能源，如风能、太阳能、水能，它考虑到满足多种需求：运输和流动、冷却、照明、定位、安全，等等。所产生的能量大部分被直接利用，如用风力推动船的前进，用冷藏的方式保持鱼的新鲜。如果风力首先用来产生电动引擎的能量，再推动船前进，那么就只剩下不到一半的能量能够被利用了。

当把可再生能源和能源效率同几何学、流体动力学和材料学相结合来减少阻力时，虽然系统变得更加复杂，但性能却得到了改善。如果我们利用当地可获得的水流和风，那么可持续发展就能取得巨大的成就。

动手能力

Capacity to Implement

机翼或帆的形状决定了飞机如何飞行，或者一艘船如何航行。选择你喜欢的3种类型的帆或者机翼，学习如何把它们画出来。然后试着解释它们如何影响一艘船或一架飞机的空气动力。不需要了解特定、精确的几何知识，看看你能不能画出空气沿船帆或机翼流动的方式，试图解释其中的原理。为什么船帆或机翼能够工作呢?

故事灵感来自

This Fable Is Inspired by

安德烈·柯尔莫哥洛夫

Andrey Kolmogorov

安德烈·柯尔莫哥洛夫生于坦波夫，距离莫斯科 500 多千米。他 5 岁时就已经开始学数学了。他毕业于莫斯科国立大学和莫斯科门捷列夫化学技术研究所，后来成为一位大学教授。他专注研究湍流，后来专注于复杂性算法理论。1956 年，他出版了一本关于概率论基础的书。他不仅是一位大学教授，也积极参与天才儿童教育学的发展。他生前是俄罗斯科学院院士，获得过众多奖项和荣誉，包括列宁奖和沃尔夫奖。

图书在版编目(CIP)数据

冈特生态童书.第三辑修订版：全36册：汉英对照 /
(比)冈特·鲍利著；(哥伦)凯瑟琳娜·巴赫绘；
何家振等译.—上海：上海远东出版社，2022
书名原文：Gunter's Fables
ISBN 978-7-5476-1850-9

Ⅰ.①冈… Ⅱ.①冈… ②凯… ③何… Ⅲ.①生态环
境-环境保护-儿童读物—汉、英 Ⅳ.①X171.1-49

中国版本图书馆CIP数据核字(2022)第163904号

著作权合同登记号图字09-2022-0637号

策　　划　张　蓉
责任编辑　祁东城
封面设计　魏　来李　廉

冈特生态童书
伴日航行
[比]冈特·鲍利　著
[哥伦]凯瑟琳娜·巴赫　绘
高　芳　李原原　译